Health 242

活得好还是活得久？

Live to be a Hundred?

Gunter Pauli

［比］冈特·鲍利 著

［哥伦］凯瑟琳娜·巴赫 绘

章里西 译

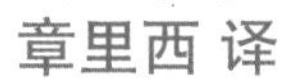

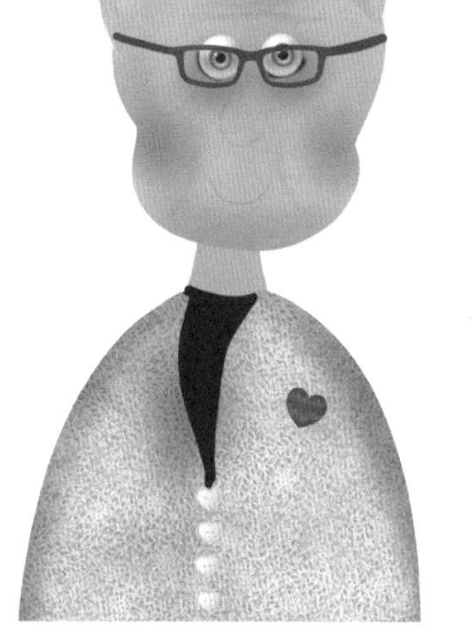

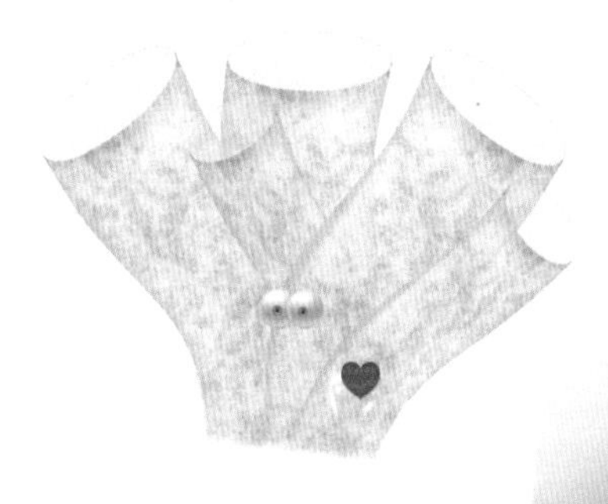

上海远东出版社

特别感谢以下热心人士对童书工作的支持：

匡志强　方　芳　宋小华　解　东　厉　云　李　婧
刘　丹　熊彩虹　罗淑怡　旷　婉　杨　荣　刘学振
何圣霖　王必斗　潘林平　熊志强　廖清州　谭燕宁
王　征　白　纯　张林霞　寿颖慧　罗　佳　傅　俊
胡海朋　白永喆　韦小宏　李　杰　欧　亮

目录

Contents

墨西哥湾来的黑珊瑚和东海来的玻璃海绵正在召开紧急会议。他们都发现，由于气候变化和海水变酸，生存变得越来越艰难。

“问题是人类燃烧了太多的化石燃料，”海绵说，“另外，为什么他们都想长寿，都想活到100岁！”

A black coral from the Gulf of Mexico and a glass sponge from the East China Sea are holding an emergency meeting. Both find that life is getting harder due to climate change and seawater turning acidic.

“The problem is that people burn too much fossil fuel,” the sponge says. “And what is it with them wanting to live longer, to up to a hundred years!”

黑珊瑚和玻璃海绵……

A black coral and a glass sponge ...

有些甚至要灭绝……

Some of us may even go extinct ...

“我听说统计数据显示，人类的确活得越来越长了，”珊瑚回应道。

“人类显然忘记了，许多动植物的寿命都比他们长得多。但是现在，由于人类挥霍的生活方式，其他生物的生命都在缩短。有些甚至要灭绝……”

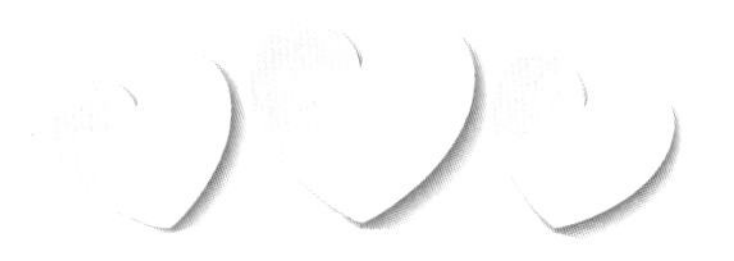

“I heard that statistics show that people are indeed living longer and longer,” the coral responds.

“What people clearly forget is that many animals and plants live much longer than they do. And now, due to their wasteful lifestyle, it seems that everybody else’s lives may be cut short. Some of us may even go extinct…”

珊瑚说：“唉，我都快2 000岁了，现在却要为生存担忧。”

“是啊，我也数不清自己的年龄了——好像很久前我就10 000岁了。真的，这是真的！有人告诉我，我是地球上最古老的动物。”

“天哪！说到年龄，我听说某些松树和柏树有差不多5 000岁了。”

“Well, I have been around for two thousand years, and am now feeling the strain of trying to survive,” Coral says.

“And I have lost count of my age – long before I reached ten thousand years. Yes, it’s true! I have been told that I am the oldest animal on Earth.”

“My goodness! Speaking of age, I know of pine and cypress trees that are nearly five thousand years old.”

我是地球上最古老的动物……

I am the oldest animal on Earth ...

……纳米布沙漠生长的千岁兰。

… Welwitschia growing in the Namib Desert.

“还有，你听说过已经茁壮成长2 000多年的猴面包树和橄榄树吧？”

“给我留下最深刻印象的，是纳米布沙漠生长了2 000年的千岁兰。”

“可是珊瑚，人类难道没有意识到，他们试图活得更长而不是更好的生存方式，正在改变地球上我们其他物种的生活，并把我们置于危险之中吗？他们好像根本不关心这些对我们的影响！”

“And have you heard of baobab and olive trees, that have thrived for more than two thousand years?”

“What I find most impressive, is the Welwitschia growing in the Namib Desert for two thousand years.”

“But Coral, do people not realise that by trying to live longer instead of better, they are changing life on Earth for everyone, putting us all at risk? They don’t seem to care about their impact on us!”

“他们并不关心我们的安危，他们这会儿还在恢复过去的物种，让在古盐中找到的细菌和在永久冻土中捕获的线虫复活。他们甚至用在冰河时代松鼠巢穴里找到的种子种了树。”

“我想知道，人类为什么只想着延年益寿而不是让生命更有活力……过一种快乐健康的生活，而不仅仅是活得长点儿。”

“是的，他们应该先照顾好自己的健康。然后还得照顾好我们星球的健康，”珊瑚说。

“Instead of caring about us, they are now bringing species back from the past, reviving bacteria captured in ancient salt, and nematodes from permafrost. They even planted trees from seeds found in an Ice Age squirrel’s cache.”

“I wonder why people, instead of wanting to add years to their life, don’t add life to their years … to lead a happy and healthy life, rather than a long one.”

“They should take care of their own health first, I agree. And then they should take good care of the health of our planet,” Coral says.

恢复过去的物种……

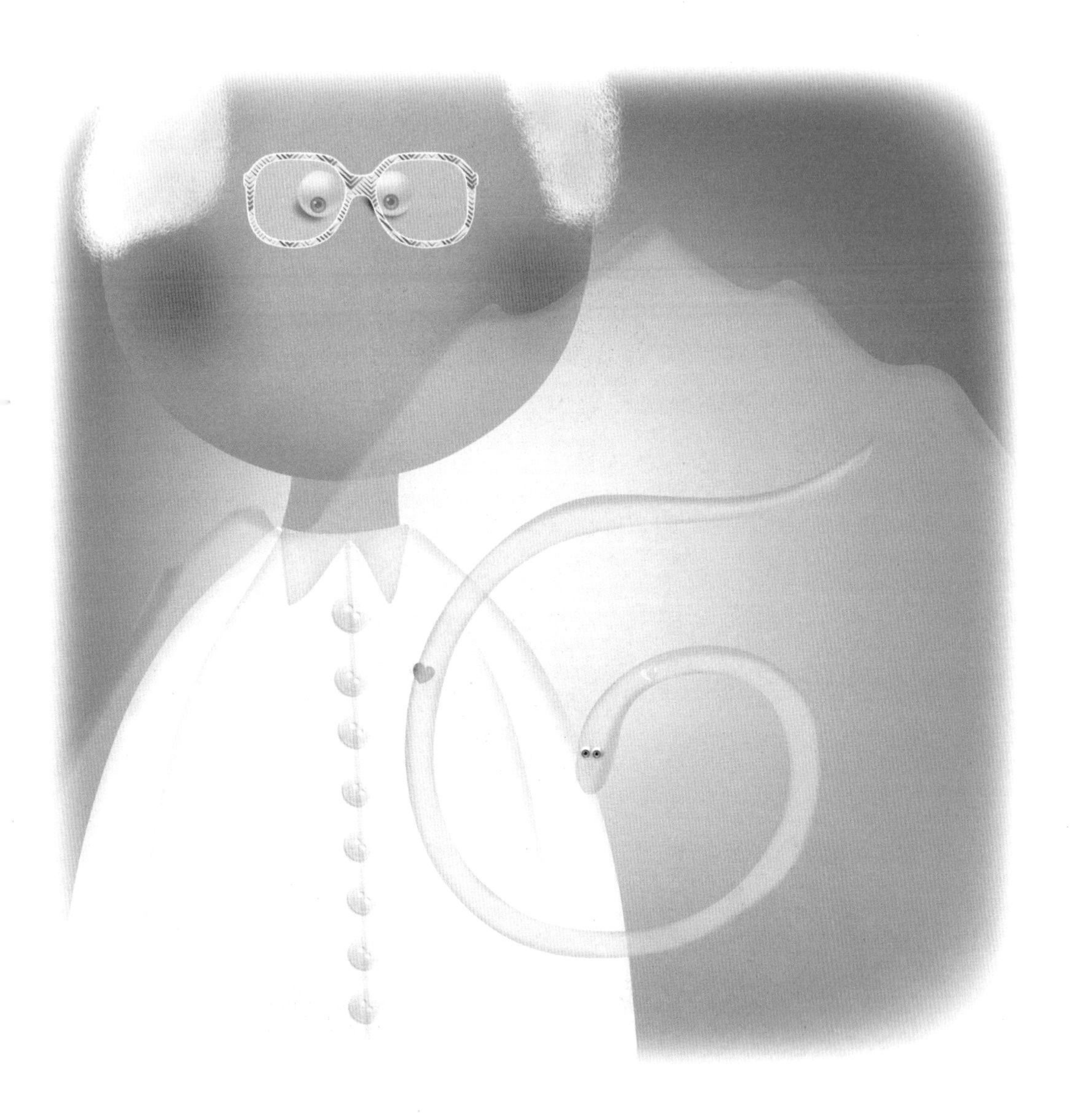

Bringing species back from the past ...

当人类开始吃海鲜……

when people started eating seafood ...

“是的，他们应该从吃当地产的应季食物开始。这肯定比吃那些绕地球半圈空运过来的化学处理过的冰冻食物要强！”

“你知道吗？当人类开始吃海鲜时，他们的大脑才开始进化。所以重要的是吃正确的食物，用正确的方式烹制——慢煮、烤、蒸、炖和大量发酵。”

“专家建议人类要保持营养，细嚼慢咽，并在进餐时进行社交活动。还要保持食物的量小；最好用各种小碟子替代大盘子。”

“Well, they could start by eating seasonal food, produced locally. Flying frozen and chemically treated food halfway around the world, I ask you!”

“Did you know it was when people started eating seafood that their brain started evolving? So important to eat the right food, prepared the right way: slow-cooking, broiling, steaming, stewing, and a lot of fermenting.”

“And experts advise people to retain nutrients, to eat slowly, to chew well – and to socialise at mealtimes. Also to keep portions small; and it is better to have a variety of small dishes instead of one big plate of something.”

“所有这些都与大多数人现在的做法背道而驰：同样的东西，想要得越来越多，一年四季周而复始，而且不管它来自世界何处。”

“他们也应该注意喝什么。富含抗氧化物的茶大有裨益。还有，当然不要吃高糖的甜点……”

“他们应该放慢节奏！由于生活节奏太快，很多人牺牲了一天中最重要的一餐：早餐。”

“我们还要提醒他们什么？要多吃海带，哦，对了，还有饭前要洗手。”

“All this runs contrary to what most are doing now: wanting more, and more of the same, and all year around too, wherever it comes from on the globe.”

“They should watch what they drink too. Tea, full of anti-oxidants, goes a long way. And then, of course, staying away from sweet desserts full of sugar…”

“And they should slow down! With people so pressed for time, many sacrifice the most important meal of the day: breakfast.”

“What else do we need to remind them of? Eat plenty of seaweed, oh yes, washing hands before every meal.”

多吃海带。

Eat plenty of seaweed.

……几乎所有最长寿的都是女性？

… nearly all of the oldest people are women?

“就像健康饮食一样，清洁也要有规范。用蒸过的棉布比那些沾有化学物质和塑料微粒的手巾好多了，因为那些东西最终会成为需要处理的海洋垃圾。”

“你知道吗？几乎所有最长寿的都是女性。”海绵问道。“看上去男性的生活确实不如女性健康。”

“我也想知道这是为什么。另外，人类应该意识到，除了吃得更好，他们还应该少吃，即使是吃好东西。”

“这很有道理。如果他们少吃点可口的食物，我们就能多点活路，地球就会有更好的生存机会！”

“Just like healthy eating, cleanliness also requires discipline. Using a steamed cotton cloth is so much better than those hand towels laced with chemicals and microplastics, that we have to deal with in the sea.”

“Do you know that nearly all of the oldest people are women?” Sponge asks. “It does seem as if men lead less healthy lives than women.”

“I wonder why. Another thing people should realise is that, apart from eating better, they should also eat less, even of the good things.”

“Now that makes so much sense. If they consume less, of the right foods, we all stand a better chance of survival, and so will the planet!”

“人类应该学会享受他们已有的生活——多呼吸新鲜空气，睡睡午觉，坐在篝火旁与家人朋友聊聊天，看看星星，好好欣赏我们生活的这个世界。”

“吃得好，活得好，让其物种也能好好生存，”海绵补充道。“看看我们的年龄，他们就会明白活到一百岁并没有什么特别的！”

……这仅仅是开始！……

“People should enjoy the life they have – smell the fresh air, take a nap when they feel like it, sit around the camp fire chatting with family and friends, and looking at the stars, appreciating the world we all live in.”

“Eat well and live well, allowing others to live well too,” Sponge adds. “Look at our age, and you’ll see there is nothing special about turning a hundred!”

... AND IT HAS ONLY JUST BEGUN!...

……这仅仅是开始！……

... AND IT HAS ONLY JUST BEGUN! ...

Did You Know?

你知道吗？

The world average age of people is 71 years. The Japanese have the highest life expectancy in the world, and in Japan Okinawans have the longest lives even though recent lifestyle changes are taking its toll.

全球人口平均寿命是71岁。日本人的平均寿命是世界上最高的，尽管近年来生活方式的改变也在敲响警钟，但日本冲绳人的寿命仍然是最长的。

Okinawans eat more tofu (fermented soy), kombu (seaweed), squid and octopus than anyone else. They cherish local vegetables like purple sweet potatoes, daikon and bitter cucumbers while drinking turmeric and jasmine tea.

冲绳居民比其他地方的人吃更多的豆腐 、海带、鱿鱼和章鱼。他们爱吃本地蔬菜，比如紫薯、白萝卜和苦黄瓜，也喝姜黄茶和茉莉花茶。

China has, for centuries, referred to Okinawa as "The Land of the Immortals". The Mediterranean diet receives more attention from media and science because there is not as much research available on the Japanese diet.

中国古代传说把冲绳称为"长生不老之地"。鉴于有关日本饮食的研究还没有那么多，所以地中海饮食受到媒体和科学界的更多关注。

One in three inhabitants of the Ikaria Island of Greece grows older than ninety. Wild mint tea is popular amongst the locals, who eat plenty of fresh vegetables from local gardens and consume large amounts of olive oil, while eating little meat or dairy.

希腊伊卡利亚岛三分之一的居民年龄超过 90 岁。野薄荷茶在当地人中很受欢迎，他们吃大量产自当地菜园的新鲜蔬菜，食用大量的橄榄油，很少吃肉或乳制品。

The Nuoro Province of Sardinia (Italy), Loma Linda in California (USA), and the Nicoya Peninsula (Costa Rica) are "blue zones" where citizens live longer. They are mostly vegetarian, eating large quantities of beans.

意大利撒丁岛的诺奥罗省、美国加州的洛马琳达和哥斯达黎加的尼科亚半岛是所谓“蓝色地带”，那里的居民寿命更长。他们大多是素食者，吃大量的豆类。

Calorie restriction, or eating less, has been a characteristic of most regions where people, who have suffered from periods of food shortage, live longer. Food scarcity seems to stall chronic disease and boost immunity.

在许多经历过粮荒时期的地区，人们已经养成限制卡路里摄入，或者少吃的习惯，而这些地区的人们通常会更长寿。食物短缺似乎可以抑制慢性疾病并增强免疫力。

一个国家的百岁老人数量与医疗费用不相关。美国医疗保健支出的占比为 16%，是日本的两倍。日本是世界上百岁老人比例最高的国家，只有 3.4% 的人口属于肥胖。

The number of centenarians in a country is not correlated with the cost of health care. The USA spends 16% on health care, double the amount spent in Japan, that has the highest proportion of centenarians in the world, and where only 3.4% of the population is obese.

长寿与饮食、睡眠和一个人的 DNA 有关。酒精是有毒性的，但适量饮酒可延年益寿。许多上了年纪的人喜欢每天喝一杯葡萄酒、雪利酒或威士忌。许多人还吃大量的野生大蒜。

Longevity is linked to diet, sleep and also one's DNA. Alcohol is toxic, but taken in moderate amounts contribute to longevity. Many of the people who grow very old enjoy a daily glass of wine, sherry or whiskey. Many also eat a lot of wild garlic.

Would you prefer a long life to a healthy and happy one?

比起健康快乐的生活，你会更喜欢长寿吗？

Does it feel right to live longer in ways that cost others their lives?

以牺牲其他物种的方式活得更久，这对吗？

What is your favourite healthy food?

你最喜欢的健康食物是什么？

Do you think you can eat less, even of the foods that are good for you?

你觉得你能少吃点吗，即使是那些对你有益的食物？

Do It Yourself! 自己动手！

Do some research on the link between a healthy diet and leading a long, happy and healthy life. Ask your friends and family if they would be prepared to change their diet if it means that they will live longer and better. Before engaging in such a discussion, make sure you have the scientific data and statistics on hand that clearly demonstrate that those people, in many parts of the world, who have a healthy diet enjoy a longer, better life. List any food items that are known to be harmful to health, and make people aware of these. Now add a list of healthy alternatives and share this with them.

研究一下健康饮食与长寿、快乐健康生活之间的关系。问问亲朋好友，为了活得更好更久，他们是否愿意改变饮食习惯。在开始讨论之前，确保手头有科学数据和统计资料能清楚地表明，在世界许多地方，健康饮食的人寿命更长、生活更美好。列出所有对健康有害的食物，让大家知晓。然后再做一份用健康食品替代的清单，与大家分享。

学科知识

Academic Knowledge

生物学	长寿和生物意义上的“永生”概念有所区别；沉睡之后的活跃；海底有几百万年历史的内岩生微生物；植物和真菌的无性系种群由个体经大量复制后形成，这种种群可以存活几百万年。
化 学	生梅子含有柠檬酸，具有碱性和抗菌作用；发酵豆汤（味噌）含有所有必需的氨基酸；纳豆发酵豆含有维生素K_1和K_2；日本的米麹甘酒含有曲酸，有助于滋润皮肤、头发和指甲；魔芋果冻含有铁、钙、磷、硒和钾；白萝卜中铁和铜的含量高；海藻富含锌、硒、碘、镁、钙、铜、维生素B_{12}。
物 理	餐盘里的食物颜色丰富，表明这是一顿健康餐。如果像彩虹一样各种颜色都有，那这顿饭就是地地道道的健康餐。
工程学	为公共卫生、劳动卫生和污染控制进行设计需要得到生物医学、土木、化学和安全工程的支撑；设计良好的供水和废水系统工程，体现健康和可持续发展理念的建筑有助于人体健康。
经济学	平均寿命、经济不平等和工业化造成的环境污染之间的相互关系；意味着减少消费和生产变得可持续的减增长与再生；“更大更一致”的增长模式是影响所有人生活的不可持续的生活方式的核心；营养学家的现代职业。
伦理学	为了活得更长，过一种没有生活质量的生活，并且所含的经济模式会剥夺其他物种的生计，可以接受吗？吃得过多反而会影响到你个人的健康。
历 史	冰河时代；近年来非洲的平均寿命因艾滋病流行而受到影响；14世纪黑死病对人类平均寿命的影响。
地 理	由于公共卫生、医疗保健、饮食、获得教育和机会均等各方面的差异，平均寿命有着极大的不同；苏格兰女性和男性的平均寿命比西欧任何地方都要低（称为格拉斯哥效应）；东海；墨西哥湾。
数 学	平均寿命是一种生物预期平均存活时间的统计度量；冈珀茨函数；算术平均；李-卡特模型中所含死亡率的演变；预测平均寿命来计算养老金；人类发展指数将预期寿命与成人识字率、教育和生活水平结合起来。
生活方式	平均寿命与健康平均寿命之间的差异；吸毒、吸烟、酗酒、肥胖、饮食和运动影响平均寿命；糖尿病使平均寿命缩短10至20年；八分饱，儒家的观念是只要吃到八分饱。
社会学	低凝聚力、低社会资本、宗教宗派主义和缺乏社会流动性导致平均寿命降低；日本学校有营养学家帮助孩子们选择食物。
心理学	疏离和悲观的文化，压力引起的身体反应，以及不良的童年经历都会导致较低的平均寿命。
系统论	空气污染增加了哮喘和癌症的发病率；全球气温上升导致传染病蔓延，这就需要设计、扩大和实施公共卫生解决方案，利用工程专业知识开发新的系统、技术和政策，以减轻环境威胁和保护人类生命；日本人的一个观念是“勿体无”，即“俭以防匮”，特别是针对食物浪费。

情感智慧

Emotional Intelligence

玻璃海绵

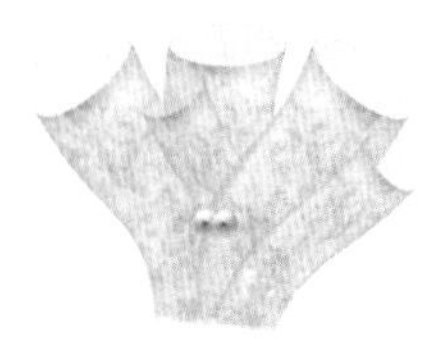

海绵关心世界的未来，指出人类是问题的根源。她解释说，人类想要延年益寿的同时往往会导致其他物种的寿命缩短。她比较了不同物种的寿命，反复指出人类肆意挥霍的生活方式。她想知道为什么人类追求的是延年益寿，而不是更好的生活，不是幸福和健康。她很务实，承认不知道所有的解决方案，但列举了改善人类生活的建议。她清楚不同性别的寿命差异，提出对所有人都有利的改革建议，并认为活到100岁并非一个值得追求的目标。

黑珊瑚

黑珊瑚与海绵的观点一致。他脚踏实地，认为长命百岁跟自己没什么关系，因为他已经2 000岁了，而且他的朋友当中还有5 000岁的。他对沙漠植物千岁兰表示钦佩。珊瑚不明白为什么人类要去恢复远古物种的生命，却不关心现存物种的生存。他感到遗憾的是，人类不仅没有照顾好自己，也没有照顾好地球的健康。他和海绵一起讨论，并提出自己的建议，在批评人类的同时，也提供了解决方案。他希望人类更好地照顾自己和地球，远离现在“更大、更远、相同的东西越来越多”的生活方式。他很清楚减增长和克制过度饮食文化的重要性，以及人们减轻压力的必要性。

艺术

The Arts

尊老敬老（儒家的价值观）和个人对长寿甚至永生的追求（道教的关怀）都反映在视觉艺术上。中国艺术中，寿桃、桃花、鹿、鹤、常青树、白色羽毛和花生（长生果）都是长寿的象征。让我们用这些象征长寿的图案绘制一幅赏心悦目的图画，作为送给80岁老人的一份生日礼物。在上色之前，记得先画一个草图。

思维拓展

Systems: Making the Connections

许多人害怕面对死亡，便去追求长寿。平均寿命是增加了，但寿命的长度等同于生活质量吗？看看许多人追求的生活方式，他们在早年以牺牲自己的健康来获得财富和地位，但后来为了健康又不得不放弃财富和地位。人类的消费模式和生活方式往往与人类自身、其他物种和地球的健康相冲突。有观点认为，如果一个人不能照顾好自己的健康，就不可能照顾好社会和生态系统。我们要采取紧急措施，改变我们的饮食，改造工业和社会，以提升所有物种的生活质量。向更快乐、更健康的生活方式转变并不难。我们不仅吃得过多，还吃错了食物。只有首先解决个人生活中的这些问题，才有可能成功解决与我们星球未来有关的更广泛的问题。这其中包括我们赖以生存的正在被无情毁灭的生物多样性。我们要从自己开始，改变我们的思维，扭转这一局面，把我们的“系统”变成一个健康、可持续发展的系统。我们要把当前的经济模式从追求标准化和高产量，转变为考虑到所有物种的需求、健康和生存。千里之行始于足下，这种转变从我们健康的早餐开始。

动手能力

Capacity to Implement

想要改变饮食习惯，可以比较一下地中海和冲绳的饮食。选几道你喜欢或者从没有尝试过的菜。这些食材和食谱都是你没见过的，指导亲朋好友如何去准备和制作。首先，列出需要的食材，比如海藻或者亚洲李子。再描述每道菜是如何烹制的。重点包括各种制作发酵食品的方法。最后，还要详细说明每种食物的营养成分。这样，大家就能知道食物的营养价值，以及对健康的影响。关键是要认识到，我们有能力科学地守护健康。我们不仅能照顾好自己，也能照顾好我们宝贵的星球。

故事灵感来自

This Fable Is Inspired by

菲奥娜·尤玛

Fiona Uyema

菲奥娜·尤玛在爱尔兰的一个牧羊场长大。她在爱尔兰都柏林城市大学学习国际市场营销和日语。在日本留学和工作期间，她从当地的亲朋好友那里接触到了日本美食。她深受丈夫吉尔玛的影响。吉尔玛生在巴西，母亲和祖父母曾在日本冲绳生活，从他们那里他学到了制作优质食品的丰富知识。冲绳以居民长寿而闻名。菲奥娜可谓是日本美食和文化大使。她写了几本有关烹饪的书，包括由 Mercier 出版社出版的《轻松制作日本料理》（2015）。菲奥娜在维珍媒体电视节目中扮演厨师。她还创立了一个名为“Fused”的亚洲优质食品品牌，该品牌保持严格的健康标准，不含糖和味精（MSG）。

图书在版编目(CIP)数据

冈特生态童书.第七辑:全36册:汉英对照 /
(比)冈特·鲍利著;(哥伦)凯瑟琳娜·巴赫绘;
何家振等译.—上海:上海远东出版社,2020
ISBN 978-7-5476-1671-0

Ⅰ.①冈… Ⅱ.①冈… ②凯… ③何… Ⅲ.①生态
环境-环境保护-儿童读物—汉英 Ⅳ.①X171.1-49

中国版本图书馆CIP数据核字(2020)第236911号

策　　划 张　蓉
责任编辑 程云琦
封面设计 魏　来李　廉

冈特生态童书
活得好还是活得久?
[比]冈特·鲍利　著
[哥伦]凯瑟琳娜·巴赫　绘
章里西　译